AMPHIBIANS

ANIMALS IN DISGUISE

Lynn Stone

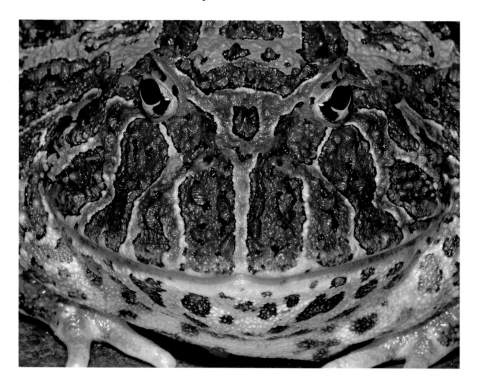

The Rourke Corporation, Inc.
Vero Beach, Florida 32964

PHOTO CREDITS
© Peter James: cover, title page, pages 10, 21; © James P. Rowan: page 18; © Lynn M. Stone: pages 4, 7, 8, 12, 13, 15, 17 and small cover photo

EDITORIAL SERVICES:
Penworthy Learning Systems

Library of Congress Cataloging-in-Publication Data

Stone, Lynn M.
     Amphibians / Lynn M. Stone.
        p. cm. — (Animals in disguise)
     Includes index
     Summary: Describes how various amphibians use ways to disguise themselves and fool other animals, including camouflage and other tricks with color and shape.
     ISBN 0-86593-487-8
     1. Amphibians—Juvenile literature. 2. Camouflage (Biology) —Juvenile literature.
[1. Amphibians. 2. Camouflage (Biology)]   I. Title II. Series. Stone, Lynn M. Animals in disguise.
QL644.2.S755 1998
597.8'1472—dc21   98–6326
                                                                                                      CIP
                                                                                                      AC

Printed in the USA

TABLE OF CONTENTS

AMPHIBIANS

Walk around a pond and you may hear a frog croak. But chances are you won't see a frog, unless it jumps. That's because frogs and other **amphibians** (am FIB ee unz)—toads, salamanders, and **caecilians** (suh SIL yunz)—wear natural disguises.

A **disguise** (dis GYZ) is a way an animal hides itself without actually disappearing. For example, the green skin of a frog makes the frog blend into the green pond plants.

It takes sharp eyes to spot a green frog sitting on a lily pad.

STAYING ALIVE

The frog disguises itself in the pond's plant life so it can stay alive. Without a disguise, the frog would be easy to see—and catch! Then it would be **prey** (PRAY), or food, for a **predator** (PRED uh tur).

Predators are animals that kill other animals for food. Amphibians are predators of the insects and little animals they catch. Amphibians are prey for snakes, fish, birds, and furry hunters.

Amphibians, like this toad, often escape being seen because of their coloring.

AMPHIBIANS IN DISGUISE

To **survive** (sur VYV), or stay alive, an amphibian has to eat, but not be eaten. In disguise, an amphibian looks like something other than what it really is. That helps an amphibian survive in a world full of predators. A disguise also helps a hunting amphibian fool its prey.

For example, a green frog in green plants doesn't frighten away its insect prey. The insects can't see the frog among the plants, so they have no fear.

A pinewoods tree frog looks like part of the tree's bark.

CAMOUFLAGE

A frog lying in green plants doesn't look like a frog to most animals. The frog's skin acts as **camouflage** (KAM uh FLAHJ).

Camouflage is an animal's ability to blend into its surroundings. A frog's coloring helps it to look like part of its background.

Toads are close cousins of the frog. Most of them are brownish, however, to blend in with their surroundings. Toads spend much of their lives on dry land.

*The Asian leaf toad looks like
the leaves on which it sits.*

The red-backed salamander wears cryptic colors for life on the forest floor.

The American toad counts on cryptic coloring to keep it safe from toad-eating predators, such as hognosed snakes.

CRYPTIC COLORS

The skin, fur, or feather colors that help an animal camouflage itself are called **cryptic** (KRIP tik) colors. Cryptic coloring is important to amphibians. Most amphibians are not armed to defend themselves in other ways.

Amphibians are generally smooth-skinned animals with moist, soft bodies. Many depend upon cryptic coloring for survival. Several kinds of tree frogs are good examples. Their colors help them blend with such objects as leaves or tree bark.

The green tree frog of the Southeast matches its green surroundings.

TRICKS WITH COLORS

Cryptic coloring allows an animal to hide fairly safely in its surroundings. If the animal changes surroundings, though, it may be seen—and attacked.

Certain amphibians avoid this problem by seeming to change their skins! Of course, they don't really change skin. They do change skin color.

Certain kinds of frogs are among these tricky little animals. When a light green frog moves to dark green surroundings, its skin darkens to match.

Many frogs change their skin color to match a change in background.

COLOR SAFETY

Common North American frogs have greenish or spotted backs. That cryptic coloring makes them hard to see for a raccoon, heron, or even a little boy or girl. Underneath, the frog has a white belly. That's a safe color, too.

For an underwater predator looking up, the pond surface is bright. The white helps the frog blend with the water surface.

The green glass frog of South America has a different kind of safe color—almost no color. You can nearly see through the frog!

The color of the glass tree frog's background shows through the animal.

SHAPES FOR DISGUISE

The toad in a woodland looks like a lump of dirt. Some of the toad's cousins have even better disguises.

A Brazilian tree frog has three fleshy "horns" over each eye. The frog's cryptic coloring and horns make it look like a fallen leaf.

A tree frog in Ecuador is shaped and colored like a bird dropping. That disguise gives the frog the look of something no predator wants to eat. Looking unfit to eat is an animal's best disguise.

This horned frog's "horns" make it look like a turned-up leaf.

SPOTS AND LINES

An amphibian's disguise sometimes depends on more than cryptic coloring or shape. Black lines around a green frog's eyes help hide the dark eyes. The black spots on the light brown skin of a leopard frog help disguise its shape.

The eyespot frog of South America has two spots on its rump. They look like huge, frightful eyes. The "big eye" disguise may be enough to confuse or even scare away a hungry snake.

Glossary

amphibian (am FIB ee un) — any one of a group of soft, moist-skinned animals that are usually born in water and usually become air-breathing land animals as adults; frogs, toads, and salamanders

caecilians (suh SIL yunz) — a group of burrowing, wormlike amphibians

camouflage (KAM uh FLAHJ) — the ability of an animal to use color, actions, and shape to blend into its surroundings

cryptic (KRIP tik) — that which helps hide, such as the colors of an animal that help it hide in its surroundings

disguise (dis GYZ) — a way of changing an animal's appearance

predator (PRED uh tur) — an animal that hunts other animals for food

prey (PRAY) — an animal that is hunted by other animals

survive (sur VYV) — to stay alive

INDEX

FURTHER READING:

Find out more about Animals in Disguise with these helpful books:
• Carter, Kyle. *What Makes an Amphibian?* Rourke, 1997
• Clarke, Barry. *Amazing Frogs and Toads*. Knopf, 1990
• Clarke, Barry. *Amphibian*. Knopf, 1993
• Greenway, Theresa. *Jungle*. Knopf, 1994
• Parker, Steve. *Natural World*. Knopf, 1994
• Stone, Lynn. *Frogs*. Rourke, 1993